Cambridge Primary

Hodder Cambridge Primary
Science
Workbook

Stage 4

Rosemary Feasey
Series editor: Deborah Herridge

HODDER
EDUCATION
AN HACHETTE UK COMPANY

Acknowledgements

The Publisher is extremely grateful to the following schools for their comments and feedback during the development of this series:
Avalon Heights World Private School, Ajman
The Oxford School, Dubai
Al Amana Private School, Sharjah
British International School, Ajman
Wesgreen International School, Sharjah
As Seeb International School, Al Khoud

Photo credits

p.49 *t* © Alchemy/Alamy Stock Photo; **p.59** *t* © Library of Congress; **p.59** *c* Georgios Kollidas/Fotolia; **p.59** *b* © Georgios Kollidas/Alamy Stock Photo.

t = top, *b* = bottom, *l* = left, *r* = right, *c* = centre

Practice test exam-style questions are written by the author.

While every effort has been made to check the instructions for practical work described in this book carefully, schools should conduct their own risk assessments in accordance with local health and safety requirements.

Every effort has been made to trace all copyright holders, but if any have been inadvertently overlooked the Publishers will be pleased to make the necessary arrangements at the first opportunity.

Although every effort has been made to ensure that website addresses are correct at time of going to press, Hodder Education cannot be held responsible for the content of any website mentioned in this book. It is sometimes possible to find a relocated web page by typing in the address of the home page for a website in the URL window of your browser.

Hachette UK's policy is to use papers that are natural, renewable and recyclable products and made from wood grown in sustainable forests. The logging and manufacturing processes are expected to conform to the environmental regulations of the country of origin.

Orders: please contact Bookpoint Ltd, 130 Milton Park, Abingdon, Oxon OX14 4SB. Telephone: (44) 01235 827720. Fax: (44) 01235 400454. Lines are open from 9.00–5.00, Monday to Saturday, with a 24 hour message answering service. You can also order through our website www.hoddereducation.com

© Rosemary Feasey 2017

Published by Hodder Education

An Hachette UK Company

Carmelite House, 50 Victoria Embankment, London EC4Y 0DZ

Impression number   9

Year   2021

Cover illustration © Steve Evans

Illustrations by Vian Oelofsen

Typeset in FS Albert 15 on 17pt by IO Publishing CC

Printed in Great Britain by Ashford Colour Press Ltd., Gosport, Hampshire

A catalogue record for this title is available from the British Library

9781471884214

Contents

Biology

Unit 1 Humans and animals	4
Self-assessment	15
Unit 2 Living things in their environment	16
Self-assessment	29

Chemistry

Unit 3 States of matter	30
Self-assessment	43

Physics

Unit 4 Magnetism	44
Self-assessment	54
Unit 5 Electricity	55
Self-assessment	65
Unit 6 Sound	66
Self-assessment	80

Unit 1 Humans and animals

Alive or not alive?

1 Which of these are alive? Which are not alive? How do you know? Give as many reasons as you can.

	Alive	Not alive	How I know
book		✔	*It does not move on its own.*
lion			
palm tree			
starfish			
plastic bottle			

2 All living things carry out life processes. Name four of the life processes. The first letter has been given.

M _____   G _____

R _____   N _____

Sorting animals

1 Below are the names of eight vertebrates (animals with backbones). Write each animal in the correct vertebrate group in the table.

| rhinoceros | cobra | fish eagle | alligator |
| sardine | desert rain frog | gecko | zebra |

Mammal	Bird	Reptile	Fish	Amphibian

2 Use these words to complete the sentences.

plants   meat and plants   meat

a A carnivore eats _____.

b A herbivore eats _____.

c An omnivore eats _____.

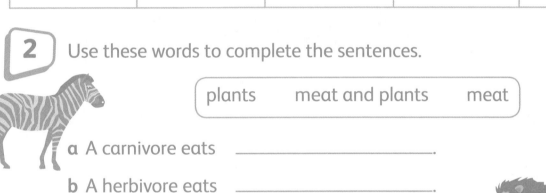

3 Complete these sentences about different types of mammals. Use the words given.

zebra   lion   human

a A _____ is a herbivore.

b A _____ is a carnivore.

c A _____ is an omnivore.

4 Challenge yourself to find out what a **detritivore** means.

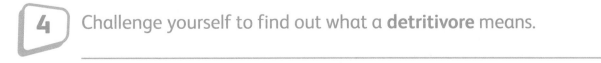

_____

_____

The skeleton

 Label the skeleton. Use these words.

| ankle | backbone | breastbone | collarbone | knee cap | pelvis |
| ribcage | ribs | shoulder blade | skull | thigh bone | wrist |

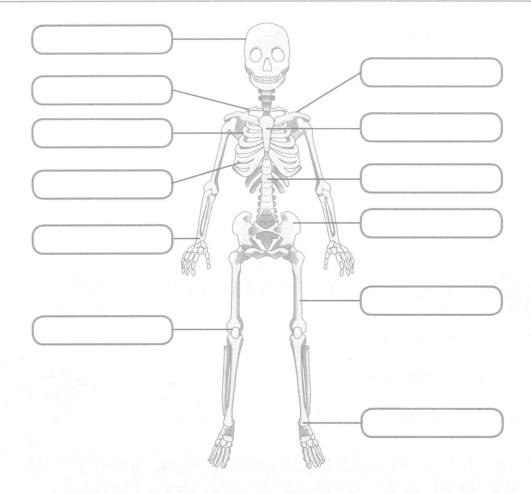

 The skeleton is important. It has three main functions (jobs). Which sentences are true? Which are false? Tick to show the correct answers.

The function of the skeleton is to …	True	False
a	support your body.		
b	hide parts of the body.		
c	protect parts of your body such as your brain, lungs and heart.		
d	help you move.		
e	keep you healthy.		

Bones

1 Complete this sentence.

Your bones are important because:

_____

_____

_____

2 Complete this sentence.

When you were a baby, you had more than _____ bones in your body.

3 Circle the foods that you should eat to help to keep your bones healthy:

- cheese
- yoghurt
- chocolate
- sardines
- okra
- sweet potato
- cake

4 Make a list of four types of exercise that you could do to keep your bones healthy.

a _____

b _____

c _____

d _____

Inside my bones

1 Complete these sentences using the words given.

> inside   hard   blood   living

Bones are _____. They grow as we grow. The outside of a bone

is very _____. Bone marrow is found _____ bones.

Bone marrow has a special job. It makes red and white blood cells.

Bone marrow is where _____ cells are made.

2 The diagram shows what a bone looks like on the inside.
Some of the words are missing. Put these words in the correct place.

> bone marrow   hard   grows

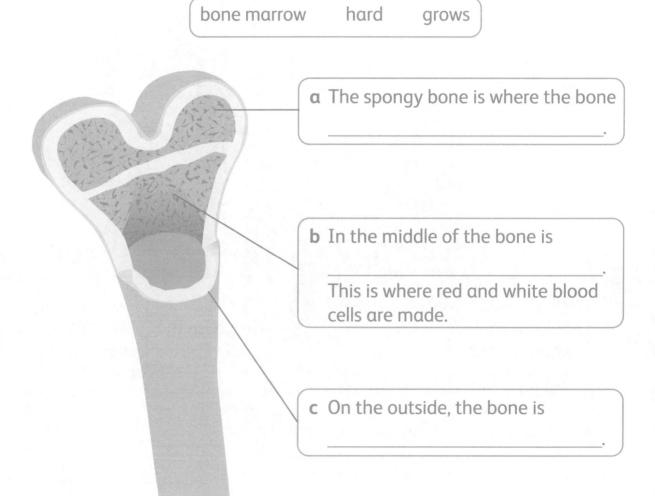

a The spongy bone is where the bone

_____.

b In the middle of the bone is

_____.

This is where red and white blood
cells are made.

c On the outside, the bone is

_____.

X-rays

 1 Poor Rianna! She was riding her bicycle when she fell off.
She hurt herself badly. She was taken to hospital.
A doctor took X-rays of her bones. Look at these X-rays.

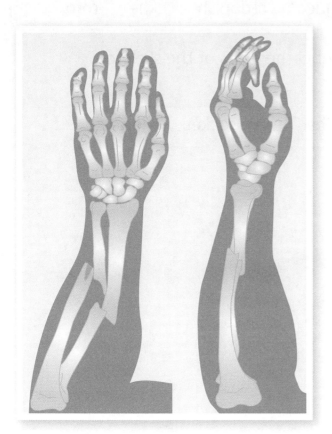

a Which part of the body can you see in this X-ray?

_____.

b Rianna has broken two bones in this part of her body. Find out the names of the bones she has broken.

_____

_____.

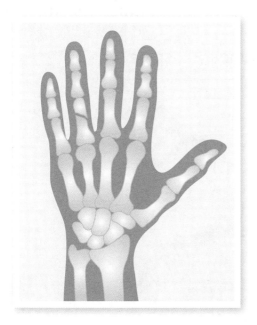

2 Rianna also broke a bone in her hand. Which finger did she break? How can you tell that the bone is broken?

_____

_____

_____

_____.

Skeleton words

1 The letters of these words are mixed up. Put them in the correct order to spell some bones in the body.

> ksull   gel   sbri   birgace   redoulsh   ekne   tofo

2 Here is the skeleton of a hawk. Label the parts of the skeleton. Use the words in the box.

> skull   feet   ribs   eye socket   leg

3 Here is the skeleton of a snake. Label the different parts. Use the words in the box.

> jaws   ribs   fangs   spine

Muscles

1 Complete the sentences using these words.

contracts   relaxes   shorter   muscles

a _____ are attached to bones.

b When a muscle _____ (gets smaller), it gets

_____. This pulls up the bone it is attached to.

c When a muscle _____, it returns to normal size.

2 Draw an outline of a body and label where other muscles are found on the human body.

Ball-and-socket joint or hinge joint?

 1 Look at the two joints.
Complete the sentences for each, using these words.

| knee | ball-and-socket | hip | hinge |

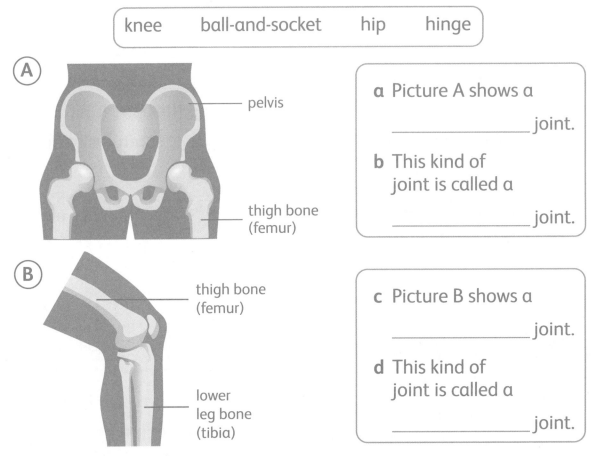

(A) pelvis

thigh bone (femur)

a Picture A shows a
_____ joint.

b This kind of joint is called a
_____ joint.

(B) thigh bone (femur)

lower leg bone (tibia)

c Picture B shows a
_____ joint.

d This kind of joint is called a
_____ joint.

 2 Draw a picture of yourself doing some exercise.
a Put a ring around different joints on your body.
b Name the joints. Are they ball-and-socket joints or hinge joints?

Medicine crossword

1 Complete this crossword. Write the answers in the correct spaces.

Across

1 This is where you should store medicines

3 These help you to get better

5 You get your medicines from this place

6 Medicines you swallow

Down

2 This is what your doctor will give you to get your medicine

4 Not all of these are medicines

2 Why do some people need to take medicine?

_____

_____

_____

3 Why do some parents lock the cupboard where medicines are kept?

_____

_____

_____

4 Why should you not take medicines that a doctor has prescribed for someone else?

_____

_____

_____

Stay safe poster

 Design and make a poster to tell other people how to be safe around medicines in the home.

STAY SAFE!

Self-assessment

Unit 1 Humans and animals

😊	I understand this well.
😐	I understand this but need more practice.
🙁	I do not understand this yet.

I need more help with …

_____

_____

_____

Learning objectives	😊	😐	🙁
I know that humans have bony skeletons inside their bodies.			
I know that some animals have bony skeletons inside their bodies.			
I can name some skeleton bones.			
I can describe how my skeleton grows as I grow.			
I can explain how a skeleton supports and protects the body.			
I can describe how skeletons have muscles attached to the bones.			
I can explain how muscles contract and work in pairs.			
I know why medicines are sometimes called drugs.			
I know why we use medicines.			

Unit 2 Living things in their environment

Environment word search

1 Names of different environments are hidden in this word search puzzle. Find these words:

> forest sea ocean lake river mountain desert polar

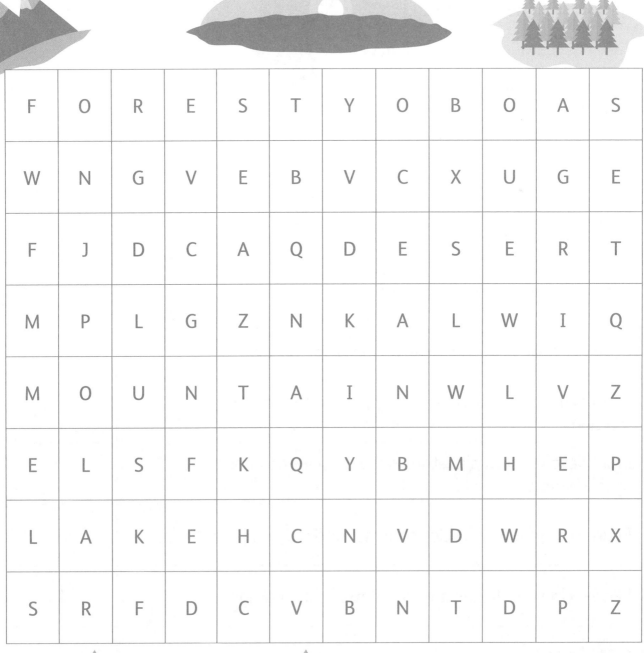

F	O	R	E	S	T	Y	O	B	O	A	S
W	N	G	V	E	B	V	C	X	U	G	E
F	J	D	C	A	Q	D	E	S	E	R	T
M	P	L	G	Z	N	K	A	L	W	I	Q
M	O	U	N	T	A	I	N	W	L	V	Z
E	L	S	F	K	Q	Y	B	M	H	E	P
L	A	K	E	H	C	N	V	D	W	R	X
S	R	F	D	C	V	B	N	T	D	P	Z

My environment

1 Draw a picture of the environment where you live.

Think about the things in your environment. These could be living and non-living things, or things built by humans.

Label the picture with the features of your environment.

The purple heron

  1 Read the fact file about the purple heron below.
Answer these questions.

a What kind of habitat does the purple heron live in?

_____

b What kind of food does the purple heron feed on in its habitat?

_____

c Where does the purple heron lay its eggs?

_____

d What will it feed its young (babies)?

_____

e Where does the purple heron shelter?

_____

Purple heron fact file

The purple heron is found in South Africa, Botswana plus many other African countries. It lives in reeds near water. It feeds on fish, rodents, frogs, reptiles and insects.

It may stalk (follow quietly) its prey. Or it may stand still, waiting to catch food as it passes by.

The colours of the purple heron help to camouflage (hide it) against the reeds in the water. It makes its nest among the reeds. Purple herons have long necks and legs.

Make your own fact file

1 Draw your favourite animal and its habitat. Research to write a fact file for that animal. You may choose an animal from any country.

_____

_____

_____

_____

_____

_____

_____

_____

2 Choose a small part of your favourite local habitat. Use a hand lens to observe the habitat carefully. Sketch your habitat. Annotate (label) your sketch with information.

a Is it in the shade or in the light?

_____

b What kind of animals and plants live there?

_____

c What food is there for the animals?

_____

d What kind of plants are there?

_____

e Is it damp or dry?

_____

f What else can you see?

_____

Rainforest habitat

 1 A rainforest is a habitat. Do you have a rainforest in your country?

    a Research rainforests. Find out seven facts about rainforests.

    b Write your facts in the shapes around the rainforest picture below.

Rainforest habitats are forests. They are found in areas close to the equator.

Classifying vertebrates

 Complete the table below. The first one has been done for you.

Animal	What kind of animal? (fish, mammal, reptile, amphibian, bird)	How I know
salmon	fish	It lives in water. It has scales. It has gills.
snake		
frog		
horse		
vulture		

Sorting animals – what do they eat?

 Here are some animal names. Sort the animals into the right places on the sorting circles. If you need to, look up herbivore, omnivore and carnivore in the scientific words dictionary in your Learner's Book.

goat	snake	crocodile	antelope
lion	snail	human	eagle
great white shark	elephant	lizard	bear
rabbit	jaguar	spider	camel

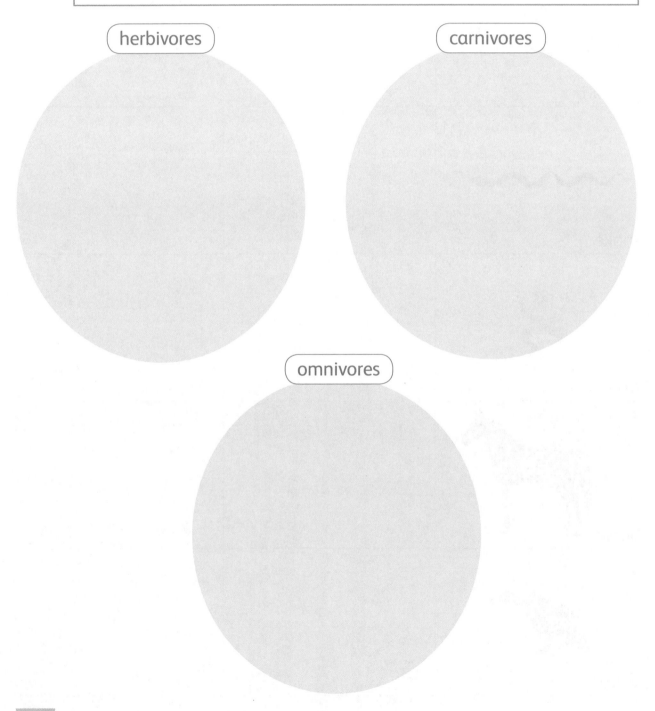

herbivores

carnivores

omnivores

Using an identification key

1 Look at the animals on this page. Write the names of the animals from the box below in the correct place on the identification key.

Is it a mammal?

Yes — No

Does it have two humps?

Is it a reptile?

Yes — No

Yes — No

It is a _____

It is an _____

It is a _____

Does it have a V-shaped nose?

Yes — No

Here's a clue: alligators have a wide U-shaped nose!

It is a _____

It is a _____

| bactrian camel | alligator | antelope | crocodile | hoopoe |

My identification key

1 Choose five different animals. Draw an identification key for these animals in the box below. Ask a partner to try it to see if it works.

Reduce, reuse and recycle

 Draw lines to match the words with the correct definitions.

(reduce)

(reuse)

(recycle)

This enables the materials you throw away to be used again by making them into new things.

Have less waste by not buying things with a lot of packaging.

Before you throw items away, think about how you can use them again. For example, you could use old jars to store buttons or pencils.

2 Write the objects from the box below into the correct bin. One has been done for you.

writing paper   container   plastic cups

newspaper   cardboard shoebox   empty yoghurt pots

glasses   glass bottle   paper bag   plastic bottle   pullover

Reduce	Reuse	Recycle
	plastic bag	

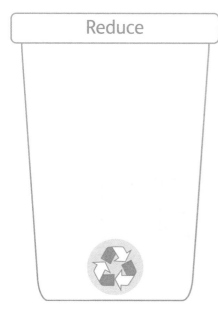

Paper plant pots

 1 Here is an idea for reusing paper. Make your own plant pots by following these instructions.

You will need...
- newspaper (black and white – not shiny or coloured)
- cardboard tube

1 Lay a full sheet of newspaper flat.

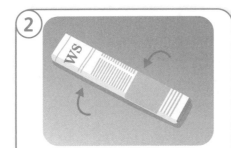

2 Fold the paper in half lengthwise twice, to make a long, narrow strip.

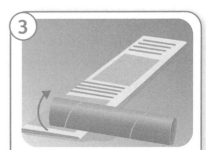

3 Lay a cardboard tube on its side. Place it on one end of the paper strip. Roll the newspaper around the tube. About half the paper strip should overlap the open end of the tube.

4 Push the ends of the paper into the open end of the tube. Do not try to be neat and tidy. Just stuff the overlapping newspaper into the tube.

5 Pull the tube out of the newspaper. You will have the newspaper pot in your hand.

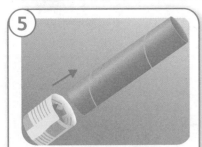

6 Put the bottom of the newspaper pot on a table. Push the tube back into the pot to flatten the folded part. This step will seal the bottom of your pot so that it is secure.

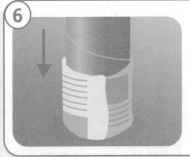

7 Remove the tube from the pot. Then fill your pot with soil. It is now ready for growing seeds!

8 A seedling is growing!

Which class uses the least paper?

Class 4 carried out a survey over three days. They wanted to find out how much paper learners threw away in Class 3, Class 4 and Class 5. Use their data on the bar chart below to answer these questions.

a Which class threw away the most blank paper? _____

b Which class threw away ten pieces of paper that had been used on both sides? _____

c Which class threw away the most paper? _____

d Which class threw away the least paper? _____

e Which class do you think wasted the most paper? _____

Think about how the paper has been used before it is thrown away. Which is the most wasteful use of the paper?

f Why do you think this class is the most wasteful?

_____

_____

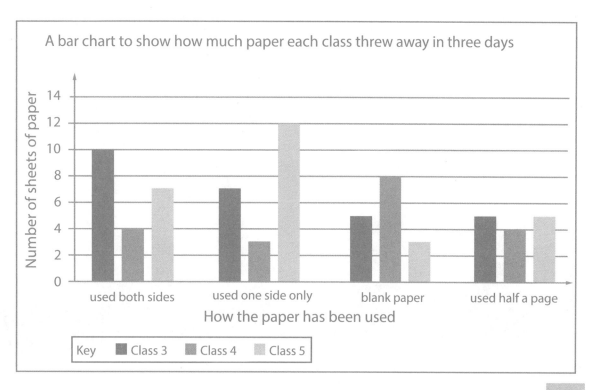

A bar chart to show how much paper each class threw away in three days

Number of sheets of paper

How the paper has been used

Key   Class 3   Class 4   Class 5

Pollution information leaflet

 Use the space below to create an information leaflet. In your leaflet, tell people about how waste plastic objects can be a danger to animals in the environment. Use some of these words in your leaflet.

animals pollution	plastic humans	eat environment	danger play	dangerous bags

Self-assessment

Unit 2 Living things in their environment

😊	I understand this well.
😐	I understand this but need more practice.
🙁	I do not understand this yet.

I need more help with …

_____

_____

_____

Learning objectives	😊	😐	🙁
I can explain what a habitat is.			
I know why different animals are found in different habitats.			
I can describe how an animal is suited to the environment where it lives.			
I can use an identification key.			
I know what reuse, recycle and reduce mean.			
I know why humans should reduce waste.			
I know what I can do to reduce waste.			
I can describe how humans pollute the environment.			
I can describe how pollution can harm some animals.			

Properties of a plastic bottle

 1 Which properties does the plastic bottle have? Put a circle around each property. Choose from the words around the bottle.

waterproof

shiny

opaque

dull

transparent

hard

flexible

magnetic

soft

rigid

 2 Ahmed made a table of silly materials. Complete the 'Better material' column of the table. Add the better materials that could be used to make each object.

Object	Silly material	Better material
tea pot	chocolate	
newspaper	concrete	
shoes	newspaper	
spoon	jelly	
trousers	wood	

Solids, liquids and gases

1 Which sentences are true and which are false?
Tick to show the correct answers.

	Sentence	True	False
a	A solid never keeps its shape.		
b	A liquid can be poured easily.		
c	A liquid is very easy to hold.		
d	Liquids cannot change their shape.		
e	Gases are usually invisible – we cannot see them.		
f	Gases do not move around and fill up spaces.		

2 Look at the pictures. Think about each one. Is it a solid, a liquid or a gas?
Complete the table.

brick   puddle   rain   apple   air in balloon

pencil   book   sand   milk   spoon

Solid	Liquid	Gas

Solids alphabet

1 Write a solid object for each letter of the alphabet.

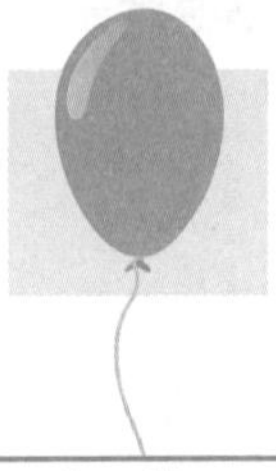

A _____

B _____

C _____

D _____

E _____

F _____

G _____

H _____

I _____

J _____

K _____

L _____

M _____

B–balloon

C–car

G–glass

H–handbag

K–key

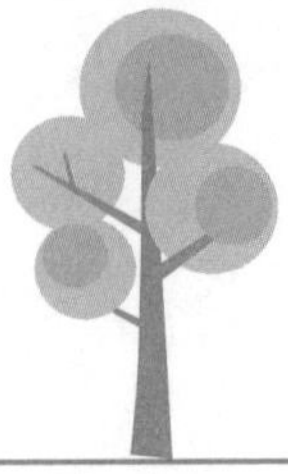

T–tree

N _____

O _____

P _____

Q _____

R _____

S _____

T _____

U _____

V _____

W _____

X _____

Y _____

Z _____

Liquids poster

 1 Think about all the liquids you have seen. Think of liquids at home and near your home.

a Make a list of all of the liquids you can think of.

b Turn your list into an interesting poster in the space below.

My poster about liquids

Tell Zoosh about solids, liquids and gases

1 Zoosh, a visitor from outer space, is very puzzled. He has no idea what solids, liquids and gases are. He needs your help.
Write in each speech bubble to explain the three states of matter.

Which liquid is the runniest?

 1 Adam and Ethan are in Class 4. They carried out a fair test to find out which liquid is the runniest. They timed different liquids as the liquids travelled from the top to the bottom of a metal tray.
Use their bar chart below to answer these questions.

   a Which liquid is the runniest? Why? _____

_____

   b Which liquid took 35 seconds to travel down the tray? _____

   c How fast did the shampoo travel? _____

   d Write down the liquids in order of how runny they are. Start with 'most runny'. End with 'least runny'.

   Most runny 1 _____

                    2 _____

                    3 _____

   Least runny 4 _____

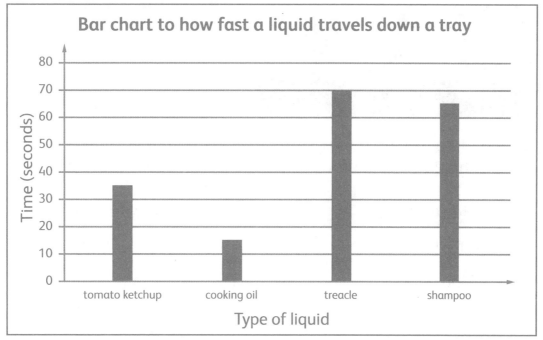

Bar chart to how fast a liquid travels down a tray

 **2** Think about the results. Why do you think the different liquids take different amounts of time to travel down the tray?
Use the word **viscosity** in your answer.

_____

Record the results

1 On page 35, Adam and Ethan carried out a fair test to find out which liquid is the runniest. They timed liquids as they travelled from the top to the bottom of a metal tray.

Adam and Ethan started to record what they did, but did not finish. Complete their notes on the fair test. Use the picture of their test below to help you.

Which liquid is the runniest?

a We changed _____

_____

_____

_____

b We used a _____ to time the liquids.

c We kept these three things the same:

1 _____

2 _____

3 _____

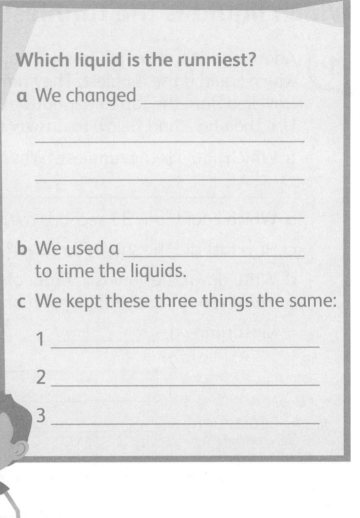

2 a Complete the table using the bar chart on page 35 to help you. Add the time it took for each liquid to run down the tray.

Liquid tested	Time taken to run down tray
tomato ketchup	
cooking oil	
treacle	
shampoo	

b Adam and Ethan repeated their fair test three times and compared the results. Why did they do this?

_____

Is there anything in the balloon?

1 a Look at this balloon.
Is anything inside the balloon?

_____

_____

_____

b What state of matter is inside the balloon?
Explain your answer.

_____

_____

_____

_____

_____

_____

_____

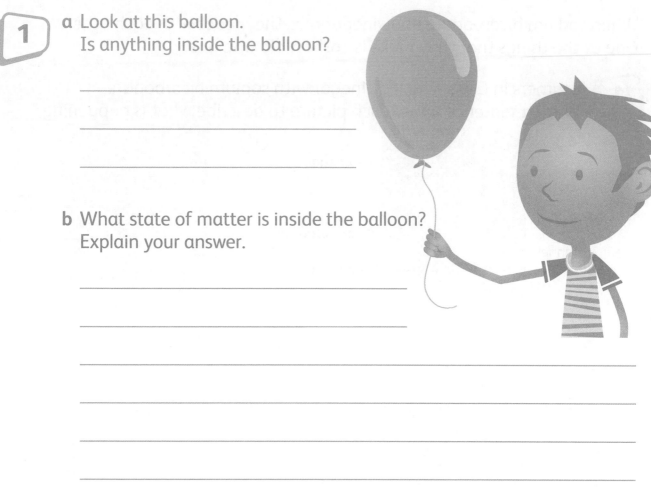

2 Read the fact about gases in each balloon. Which facts are right? Which
facts are wrong? Write BURST under the balloons with the wrong facts.

a Gases are often invisible.

b Gases cannot be squashed.

c Gases are everywhere.

d Gases cannot spread out.

_____   _____   _____   _____

Making a gas

When sodium bicarbonate and vinegar mix, they **react** with each other.
One of the things that they make is **carbon dioxide gas**.

 Learners in Class 4 mixed vinegar with sodium bicarbonate.
Write a sentence under each picture to describe what is happening.

(a)

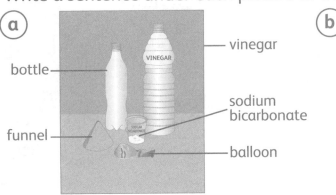

(b)

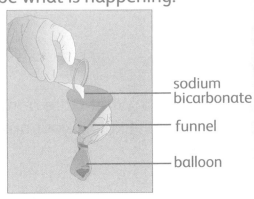

_____

_____

(c)

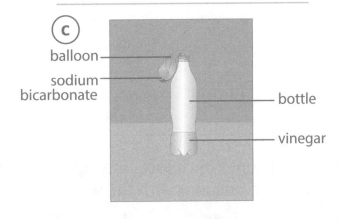

(d)

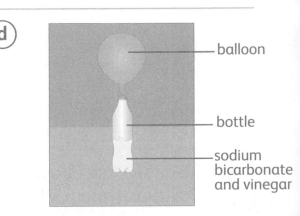

_____

_____

 Class 4 used their knowledge of mixing vinegar
and sodium bicarbonate to make a model
volcano with lava coming out of the top.

Write a set of instructions to describe
how to make the volcano.
Use a separate sheet of paper.
You could use pictures and sentences
and then make your own volcano at home.

Changing state

 1 What is happening in this diagram? Use these words:

ice	solid	liquid	frozen	melt

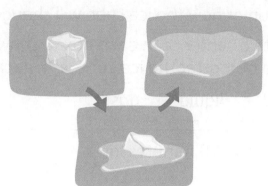

_____

_____

_____

_____

_____

2 What is happening in this diagram? Use these words:

ice	solid	liquid	freeze

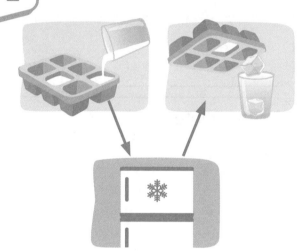

_____

_____

_____

_____

_____

3 What is happening in this diagram? Use these words:

liquid	boiling water	steam
water vapour	condensing	gas

_____

_____

_____

_____

What happens to ice?

 Try this test: Leave an ice cube on a table in your classroom for 24 hours.

  a Predict what you think will happen to the ice cube. Draw a picture in each box to show your prediction of how the ice cube will change.

  b Do the test over two days. Write a sentence in each box to show the actual results. Write what state of matter the ice cube is, in each box.

solid   liquid   gas

1 1 minute later

2 10 minutes later

3 30 minutes later

4 1 hour later

5 2 hours later

6 24 hours later

Melting points

 1 Use these words to complete the sentences:

> reversible   liquid   heated   cools   solid   melt   solidified

When some materials are heated, they _____ (become

_____). For materials to melt, they must be _____.

As they cool down again, they become _____. We say

that the liquid has _____. This is called a _____

change. The solid melts and then _____ down, and is

changed back again to a solid.

 2 Amara and Ebele found out that different materials melt at different
temperatures. They did some research. Here is their table of results.

Melting points of everyday materials	
Substance	Temperature (°C)
candle wax	60
chocolate	35
glass	1400
gold	1336
ice	0
salt	800
silver coin	879
sugar	180

Look at the table of melting points of everyday materials.
Use the information to answer these questions.

a Which material has the highest melting point? _____

b Which material has the lowest melting point? _____

c Which materials could be melted safely at school or at home?

_____

d Which materials require the most heat to melt? _____

e If the temperature in a room is around 20 °C, which material would

melt in the room? _____

The story of chocolate

 Read the story of chocolate and look at the pictures.

 The story of chocolate begins with cocoa. Cocoa grows on trees in countries near the equator.

 Inside each cocoa pod are cocoa beans. The cocoa beans are taken out and left to dry in sunlight.

③ The beans are broken up to get to the cocoa 'nibs' inside. The nibs are ground to form a paste. Other ingredients are added to the paste. This is heated and melted to form liquid chocolate.

 The liquid chocolate is poured into moulds. It is cooled. Bars of chocolate, chocolate drops and other types of chocolates are formed!

2 Use the information to answer these questions.

a Are the cocoa beans a solid, liquid or gas? _____

b Do you think the chocolate paste will be a solid or liquid? Why?

_____

c When the chocolate is poured into moulds, which state is it – solid, liquid or gas? _____

d If chocolate melts at 35 °C, what temperature would it have to be for the liquid chocolate to change state to a solid? _____

e When you eat a bar of chocolate is it a solid, liquid or gas? _____

f What would you have to do to change a bar of chocolate to a liquid?

_____

Is this a reversible change? _____

Self-assessment

Unit 3 States of matter

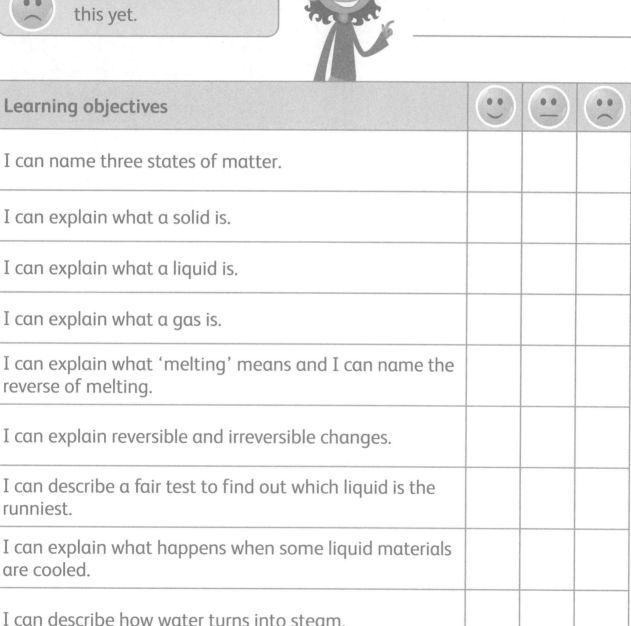

:) I understand this well.

:| I understand this but need more practice.

:(I do not understand this yet.

I need more help with …

_____

_____

_____

Learning objectives	:)	:|	:(
I can name three states of matter.			
I can explain what a solid is.			
I can explain what a liquid is.			
I can explain what a gas is.			
I can explain what 'melting' means and I can name the reverse of melting.			
I can explain reversible and irreversible changes.			
I can describe a fair test to find out which liquid is the runniest.			
I can explain what happens when some liquid materials are cooled.			
I can describe how water turns into steam.			
I can explain how steam turns back into water.			

Magnet mind map

 1 Create a mind map to show what you know about magnets. Here are some words you could use. Make sure that you write the link on the line to show why you are using the word.

attract   magnetic   metals   materials

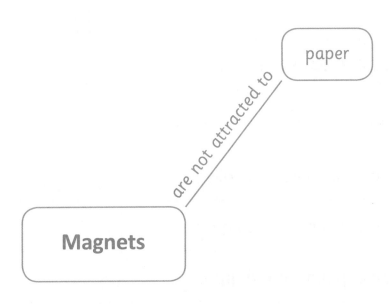

Add to this mind map throughout the unit and keep adding to it. When you learn something new, use different-coloured pencils or pens.

Attracted to a magnet

1 Look at these pictures. Think about whether each object would be attracted or not attracted to a magnet. Write the name of each object in the correct column of the table below.

ball

iron screw

brick

book

stainless steel knife

steel safety pin

nickel keyring

sea shell

shoe

plastic button

scarf

glass marble

aluminium spoon

Attracted	Not attracted

2 Think of some other objects at school or at home. Would they be attracted or not attracted to a magnet? Why? Add them to the table.

Strongest magnet

 Lami and Kamka carried out a fair test to find out how many paperclips a magnet could hold. They wanted to know which magnet was the strongest. Look at the table of their results.

Use the results to complete the bar chart below.

Remember to:
- Label the *y*-axis and the *x*-axis.
- Give your bar chart a title.
- Make sure the bars do not touch.

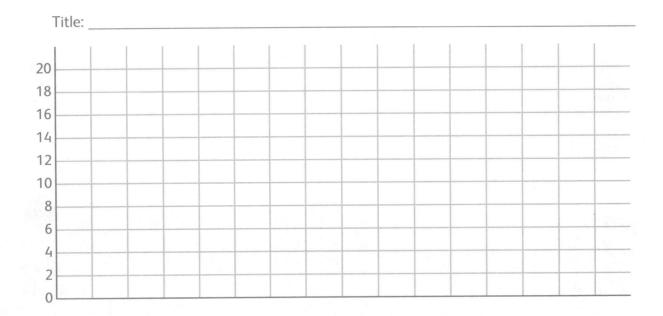

Type of magnet		Number of paperclips
circular		6
cylinder		18
horseshoe		2
bar		8
ring		10
U-shaped		3

Title: _____

 Use your graph to answer these questions.

a Which magnet was the strongest? How do you know? _____

_____

b Which was the weakest magnet? How do you know? _____

_____

More about the strongest magnet

 Sanura and Dalia tried the test on page 46. They counted the number of paperclips each magnet could hold, and recorded the results in this table.

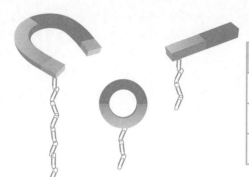

Type of magnet	Number of paperclips
horseshoe	8
bar	4
ring	3

 Ben and Eman carried out a different test. They used a ruler to measure (in centimetres) the distance from which each magnet could attract a paperclip.

Type of magnet	Distance (cm)
horseshoe	3
bar	2
ring	6

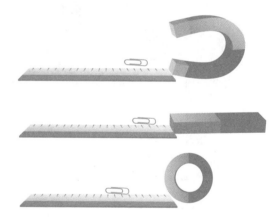

Look at the pictures and their table of results.

a Which test do you think was best? Explain your thinking.

_____

b Why do you think Ben and Eman did not get the same result about the strongest magnet as Sanura and Dalia?

_____

c Why do scientists use standard measurements such as centimetres? Why would they not record results using things like paperclips (called non-standard measurements)?

_____

Do magnets attract all metals?

1 Hari and Mitra have found out that not all metals are magnetic. They know that iron, steel and nickel metals are magnetic. Use this information to sort these objects into the correct circle. First, do some research to find out what metal each object is made from (or whether the metal is magnetic or not).

magnetic

non-magnetic

drinks can

can of food

hammer

screw

tin foil

gold necklace

copper wire

2 Find your way through this maze. Use a magnet under this page and a small object made from a material that is attracted to a magnet on top of this page. Work out the way from START to FINISH with your magnet and object. Then use a pencil to draw a way through the maze.

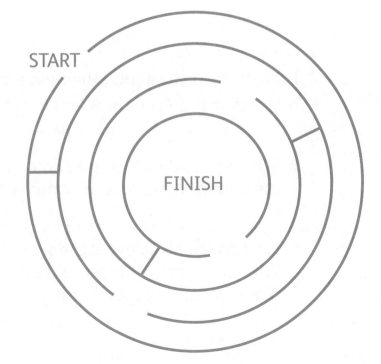

START

FINISH

Magnetic field

 Magnets have an invisible force called a magnetic field. We cannot see the magnetic field. We can see its effects if we use iron filings. Iron filings show how far the force of a magnet goes out. They also show the pattern of the force field.

Look at this picture of a magnet and its magnetic force field.

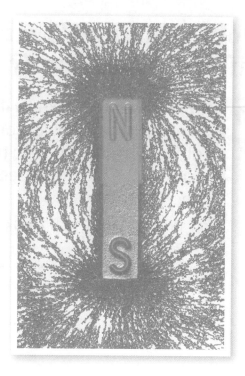

a Where do you think the force field is strongest?

_____

b How can you tell?

_____

c Where do you think the force field is weakest?

_____

d How can you tell?

_____

 Look at these magnets.

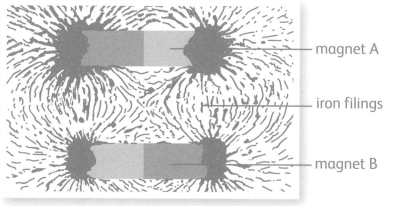

magnet A

iron filings

magnet B

Hint: Look at the magnetic field. Will the strongest magnet have more or less iron filings attracted to it?

a Which magnet do you think is strongest? _____

b How do you know?

_____

Does a magnet's magnetic field work through different materials?

 Jessica wanted to know if a magnet's magnetic field would work through different materials such as water. Look at the picture to see how she decided to find out.

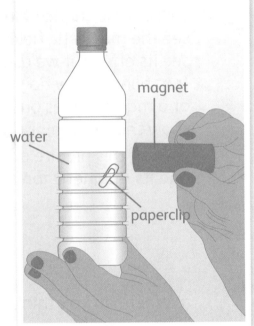

a Predict what you think will happen.

_____

_____

b Explain why you think your prediction is right.

_____

_____

 Think of a different way for Jessica to find out if a magnet's magnetic field will work through different materials. How will you make the test fair? Draw a diagram of your fair test here. Label your diagram.

North and south poles

 Zak has been learning about north and south poles on a magnet.
He knows that poles can attract – pull towards each other.
He also knows that poles can repel – push each other away.

Zak made up the words in the box. They help him to remember which poles attract each other and which poles repel each other.

> Like poles repel.   Unlike poles attract.

Use Zak's words to work out if these pairs of magnets will attract or repel each other. Tick the correct word for each pair.

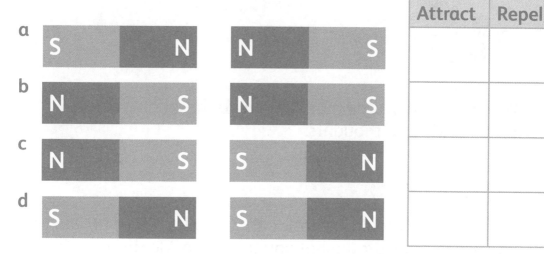

	Attract	Repel
a		
b		
c		
d		

 Zak was exploring magnets. He wondered: What will happen if I balance a magnet on a ball (cut in half) and move another magnet towards it?

Describe what Zak would have to do to make the magnet move around on top of the ball.

_____

_____

_____

Solve a puzzle

 1 These marbles are very special.
- Sometimes they can pick up other marbles.
- Sometimes they can pick up other objects.
- Sometimes they can pull along other marbles.
- Sometimes they push away other marbles.

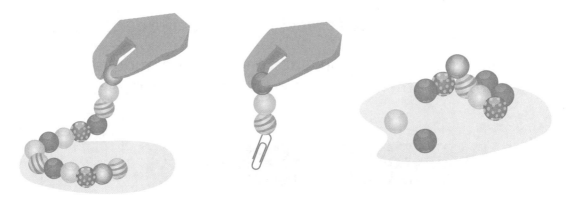

a Why do you think the marbles can do these things?
Write down your thoughts.

_____

_____

b Imagine if you could see inside the marbles. What do you think
you would find? Draw what you think is inside this marble.
Label your diagram.

Useful magnets

1 Explain how you think this kitchen utensil holder works.

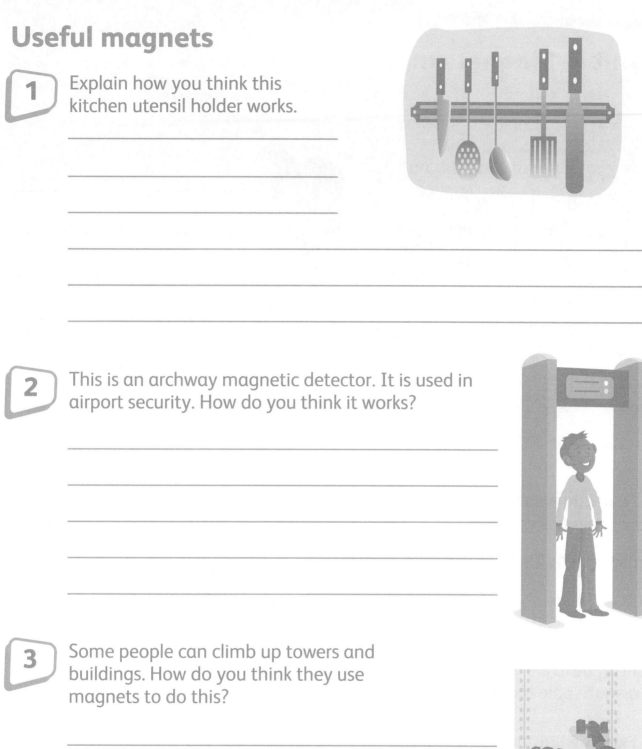

_____

_____

_____

_____

_____

_____

2 This is an archway magnetic detector. It is used in airport security. How do you think it works?

_____

_____

_____

_____

_____

3 Some people can climb up towers and buildings. How do you think they use magnets to do this?

_____

_____

_____

_____

_____

_____

Self-assessment

Unit 4 Magnetism

	I understand this well.
	I understand this but need more practice.
	I do not understand this yet.

I need more help with …

_____

_____

_____

Learning objectives	😊	😐	🙁
I can describe what a magnet is.			
I can name some metals that are magnetic.			
I can name some materials that magnets do not attract.			
I know what a magnetic force field is.			
I can name the two poles on a magnet.			
I can describe what happens if you put a north and a south pole together on a pair of magnets.			
I know what 'attract' means.			
I know what 'repel' means.			
I can describe a fair test to find out if a magnetic field will work through different materials.			
I can describe what happens if you put the same poles next to each other on a pair of magnets.			

Unit 5 Electricity

Electricity word page

1 Make an 'electricity word page' below. Draw each component (thing) in this box. Write the name of the component next to it.

wires	lamp	lamp holder	switch
motor	buzzer	cells (batteries)	

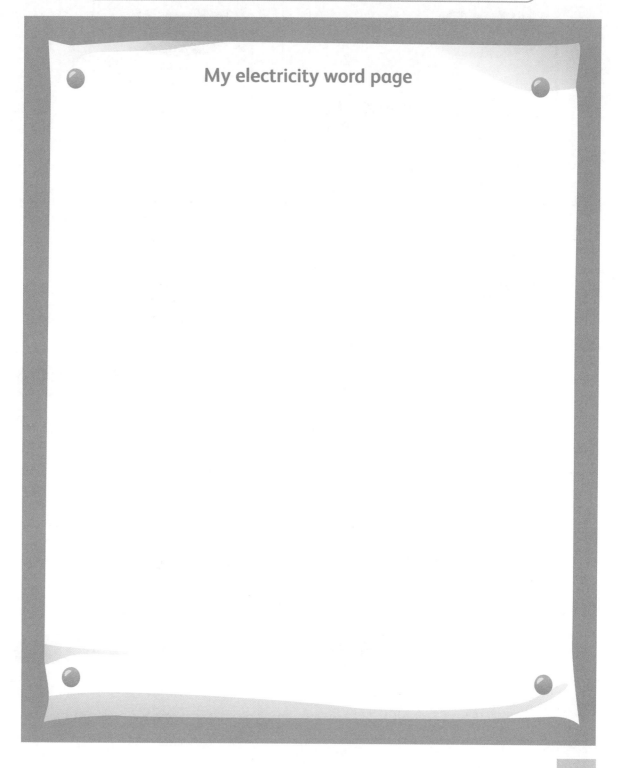

My electricity word page

Which circuit works?

1 Look at these circuits. Which ones will work to light the lamp?
Draw a circle around YES or NO.

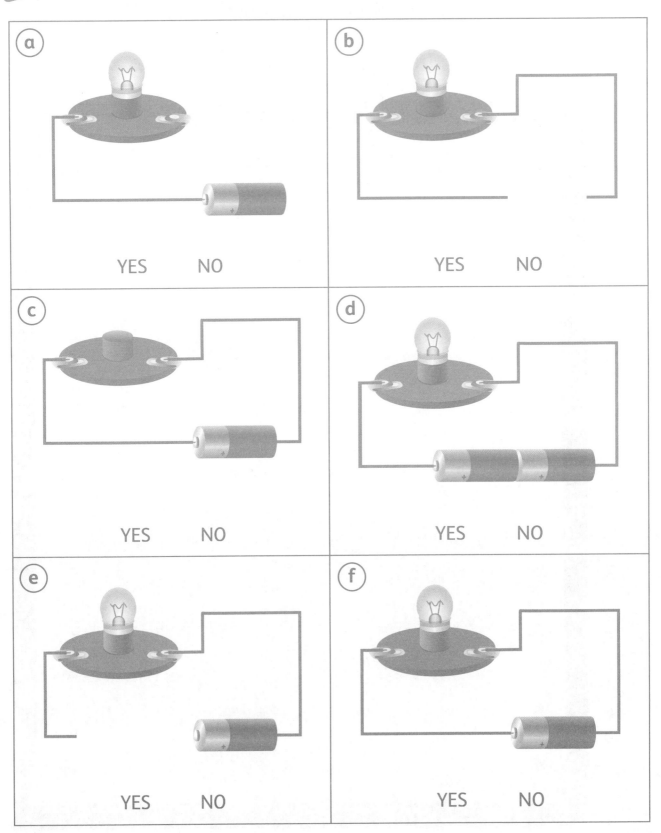

a

YES   NO

b

YES   NO

c

YES   NO

d

YES   NO

e

YES   NO

f

YES   NO

Using correct words

 1 Complete the sentences using these words.

> complete   circuit   flow   cell   break   light

a We can use electricity from a _____ to make a lamp light.

b Electricity only flows around a circuit when there is a _____ circuit.

c When there is a complete circuit, the lamp will _____.

d When there is a gap in the circuit, we say there is a _____ in the circuit.

e If there is a break in the circuit or parts do not touch properly, the electricity will not _____ around the circuit.

f When there is a break in the _____, the lamp will not light.

 2 Zaid and Ahmed raced to see who could construct their circuit in the fastest time. They measured the time in seconds.
Here are their results. They forgot to work out the average.

a Work out the average for Zaid and Ahmed. Write your answer in the last column.

Name	First time	Second time	Third time	Average time
Zaid	30	25	40	
Ahmed	35	30	25	

b Who won? _____

Hint: To find the average, add the seconds in each row. Divide by the total number of times per person (three). So, the average for Zaid is: 30 + 25 + 40 divided by 3.

Circuits

 Look at the circuit next to each letter. Complete each sentence.

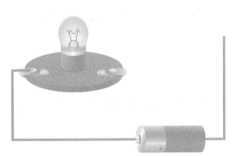

a The lamp will _____

because _____

_____

_____ .

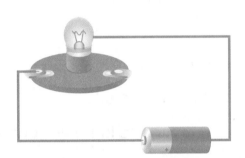

b The lamp will _____

because _____

_____

_____ .

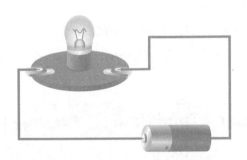

c The lamp will _____

because _____

_____

_____ .

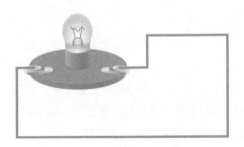

d The lamp will _____

because _____

_____

_____ .

Scientists and electricity

1 Read about these scientists. They invented things to do with electricity. Use the information to answer the questions below.

Thomas Edison
1847–1931

Thomas Edison was the first person to make a long-lasting electric lamp in his laboratory.

Benjamin Franklin
1706–1790

Benjamin Franklin was the first person to show that lightning is electricity.

Humphry Davy
1778–1829

The first electric light was made in 1800 by Humphry Davy, an English scientist.

a Who made the first electric light? _____

b What electrical component did Thomas Edison invent?

_____

_____

c What did Benjamin Franklin find out? _____

_____

_____

Cells and batteries

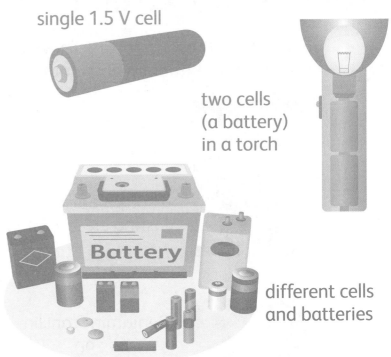

single 1.5 V cell

two cells
(a battery)
in a torch

Battery

different cells
and batteries

1 Things that use cells and batteries are called electrical appliances. Some appliances need more power than a single 1.5 V cell to work. These appliances use several cells. Two or more 1.5 V cells together are called a battery.
Cells and batteries come in different shapes and sizes. Car batteries have several cells inside them.

a Make a list of eight electrical appliances that use single cells and batteries.

_____   _____   _____

_____   _____   _____

_____   _____

b Which electrical appliances in your list do you think use the smallest cell or battery? Do some research to find out if you are correct.

_____

_____

2 Alessandro Giuseppe Antonio Anastasio Volta (1745–1827) was an Italian inventor of cells (batteries).
Look at the picture of his first cell.

Do some research. Find out four more facts about Alessandro Volta and write them here.

_____

_____

_____

_____

Circuit pictures

 Some Class 4 learners made pictures with light-up lamps. Look at Zara's picture.

Think about the circuit Zara made. Draw a diagram of the circuit she made in the box below. Label the diagram with these words:

cell	wires
lamp	lamp holder

nose lights up

 Zara would like to put a switch in her circuit. Draw the circuit picture again – this time, with a switch.

Lamps in a circuit

1 Class 4 did a test to find out what would happen if they kept adding more lamps to their circuit.

Here are their results. Some are missing. Fill in the missing results.

Number of lamps	What happens in the circuit?
1	*Lamp is very bright*
2	
3	*Lamps are dimmer*
4	
5	

2 Complete each line of the word 'electric' below with a word that relates to (has to do with) electricity. One letter has been done for you.

E
L
E
C
T
R
I *Insulator: a material that does not let electricity pass through*
C

Making electricity

 Look at the pictures. They all have something to do with electricity. How? Do some research and write your answers below.

a _____

_____

_____

_____

_____

b _____

_____

_____

_____

c _____

_____

_____

_____

_____

d _____

_____

_____

_____

_____

Modelling electricity

1 A group of learners modelled what happens in a circuit. Think about what is happening in the picture to help you to answer these questions.

a Why must the learners make sure that they are all holding hands in the model circuit?

_____

_____

b What happens to the flow of electricity when the learners do not hold hands and the model circuit is broken?

_____

_____

c How could you model a switch in the circuit?

_____

_____

2 Complete the sentences using these words.

> complete   negative (–)   components   cell (battery)   break   switch

a For electricity to flow in a circuit, it always needs a power source such as a _____.

b The wires need to be connected to both the positive (+) and _____ ends of a cell (battery).

c You can put electrical _____ in a circuit (such as lamps, motors and buzzers).

d Electricity will only travel around a circuit that is _____.

e If there is a _____ in the circuit, electricity will not flow and the lamp or other components will not work.

f A _____ is a break in a circuit, which we can use to turn a lamp on and off.

Self-assessment

Unit 5 Electricity

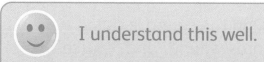 I understand this well.

  I understand this but need more practice.

  I do not understand this yet.

I need more help with …

_____

_____

_____

Learning objectives	🙂	😐	🙁
I can explain the uses of cells (batteries).			
I can make a circuit so that a lamp lights.			
I can make other components work in a circuit (such as a motor and a buzzer).			
I can explain why a component will not work if there is break in a circuit.			
I can say how to be safe around electricity in my home and outdoors.			
I can say what happens if more than one lamp is put in a circuit with only one 1.5 V cell.			
I can explain how electricity flows (moves around) in a circuit using a model.			
I can make a switch and use it in a circuit to turn a component on and off.			
I can use what I know about electricity to make a game using electrical components.			

Unit 6 Sound

Sounds around us

1 You may remember that the object making the sound is called the source of the sound. Here are the words 'sound' and 'source'. Write a source of sound beginning with each letter.

S _____

O _____

U _____

N _____

D _____

S _____

O _____

U _____

R _____

C _____

E _____

2 Damola and Jari carried out a test to find out what would happen when they walked away from the source of a sound.
They recorded their results in this table.

Distance from sound (metres)	Loudness of the sound
10	very loud
20	loud
30	quite loud
40	soft
50	not easy to hear
60	could only just hear it
70	could not hear it
80	could not hear it

Look at their results. What do you think was their conclusion? Complete these sentences.

a The further away from the source of the sound _____

_____

b The closer you are to the source of the sound _____

_____

Vibrations

 1 Complete the sentences using these words:

> vibrating   vibrations   vibrate

To make a sound, something has to _____. When you hear a sound it means that the source of the sound must be _____, even if you cannot see the _____.

 2 Class 4 have been learning about sound. They did this activity using a table tennis ball and a tuning fork.

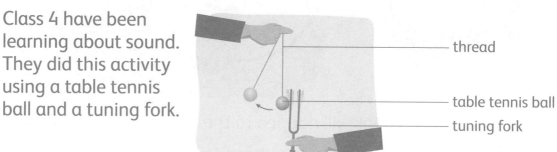

thread

table tennis ball

tuning fork

a Think about what is happening in the picture.

b Write a sentence to explain how you know that the tuning fork is making a sound. Use these words:

> sound   tuning fork   vibrations   ball

_____

_____

_____

3 Class 4 did another activity using a container filled with water and a tuning fork.

Think about what is happening in the picture. Write a sentence to explain how you know that the tuning fork is making a sound. Use these words:

> sound   tuning fork   vibrations   water

_____

_____

_____

Loud and soft

 When a sound is made, something vibrates. Think about what is happening in these pictures. Then answer the questions.

a What do you think will happen to the bird seed when you hit the drum?

_____

b What do you think will happen to the bird seed if you hit the drum very gently?

_____

c What do you think will happen to the bird seed if you hit the drum very hard?

_____

d What do you think will happen to the sound if you hit the drum very gently?

_____

e What do you think will happen to the sound if you hit the drum very hard?

_____

 Complete the sentences using these words:

> louder   softer

a The bigger the vibration the _____ the sound.

b The smaller the vibration the _____ the sound.

How you hear

 Write the letter of the correct sentence under each picture below, to show how you hear.

a The ear sends signals to the brain.

b The vibrations are felt inside the ear.

c When you hit the drum, the drum skin vibrates.

d The vibrations spread away from the source of the sound.

e The brain works out that you have heard the drum.

f When the drum skin vibrates, it makes the air around it vibrate.

_____

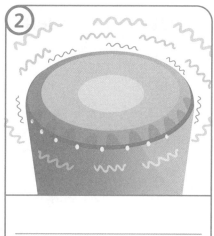

_____

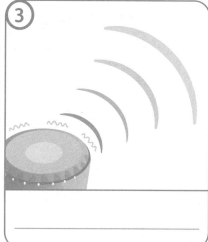

_____

_____

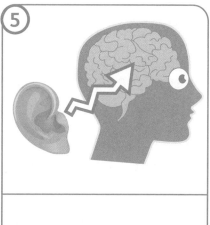

_____

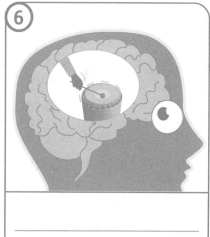

_____

Measuring sound

 Complete the sentence below using one of these words.

> millimetres   degrees   litres   decibels

Sound is measured in _____.

 Did you know that sounds over 85 dB can damage your ears and cause deafness? The table shows how loud sounds are when measured using dB (decibels). Use it to answer this question.

Which three things in this table could cause hearing loss if you were very close to them?

Source of sound	dB (decibels)	Volume
leaves rustling	10	soft
watch ticking	20	soft
someone whispering	30	soft
refrigerator	50	loud
car engine	70	loud
hairdryer	90	loud
pneumatic drill	100	painful and can damage hearing, you need ear defenders
plane taking off	120	painful and can damage hearing, you need ear defenders
space rocket launching	180	will make you deaf, you should not be near it

 The loudest human scream measured 129 dB.

a Look at the table again. Which source of sound is louder than the loudest scream? _____

b How many dB does a person who is whispering measure?

_____

c Which source of sound is softer than 20 dB?

_____

Reading a sound graph

 Learners in Class 4 left a sound level meter on in their classroom. The times were from 3 p.m. when school finished, until 9:30 a.m. the next day when the learners returned to the classroom.

This graph shows the levels of sound in their classroom.

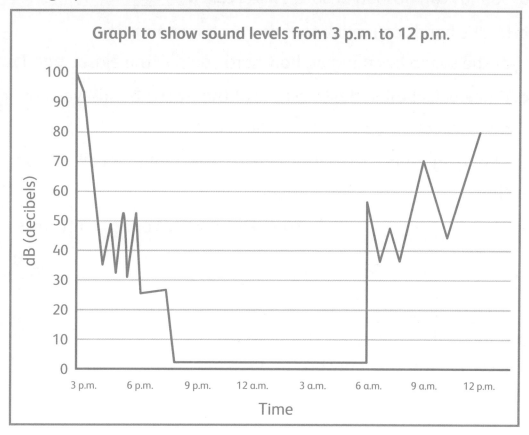

Use the graph to answer these questions.

a When were the sounds at the school the softest?

_____

b How many decibels was the loudest time in the school?

_____

c Who do you think was in the school at 6 a.m.?

_____

d What do you think happened at 9 a.m.?

_____

e What time do you think the cleaner kept switching a vacuum cleaner on and off?

_____

Changing sounds

1 Use these words to complete the sentences:

> pitch   volume   low

a Sounds can be loud or soft. This is called _____.

b Elastic bands can make sounds. You can change the _____ of the sound by changing how hard you pull the elastic bands.

c When you change the thickness of the elastic bands, you change how high or _____ the sound is.

d Changing sounds to make them lower or higher is called changing the _____ of a sound.

e The _____ of a sound is how loud or how soft the sound is.

2 Look at the guitar Sammy made.
He used elastic bands as guitar strings.
Think about what Sammy must do to change
the sound that the guitar strings make.

a Describe what Sammy has to do to make the sound louder.

_____

_____

b Describe what Sammy has to do to make the sound softer.

_____

_____

c All the elastic bands that Sammy used are the same thickness. What could he do to change the pitch of the sound?

_____

_____

d What other way could Sammy change the pitch of the sound made by the guitar strings?

_____

_____

What vibrates?

 Complete these sentences using these words:

> air   string   skin

a In a drum, the _____ vibrates to make a sound.

b In a guitar, the _____ vibrates to make a sound.

c In pan pipes, the _____ vibrates to make a sound.

2 John made some pan pipes using plastic straws and card.

a On the picture, label which straws would make notes with the highest pitch.

b On the picture, label which straws would make notes with the lowest pitch.

c Describe how John would make

a loud sound. _____

_____

d Describe how John would make a soft sound.

_____

_____

e Describe how John would make a loud, high-pitched sound.

_____

_____

f Describe how John would make a soft, low sound.

_____

_____

Spoon gong

1 Look at Imari tapping a spoon against a table.

string
vibrates
spoon
metal
ears
hit
travels
sound

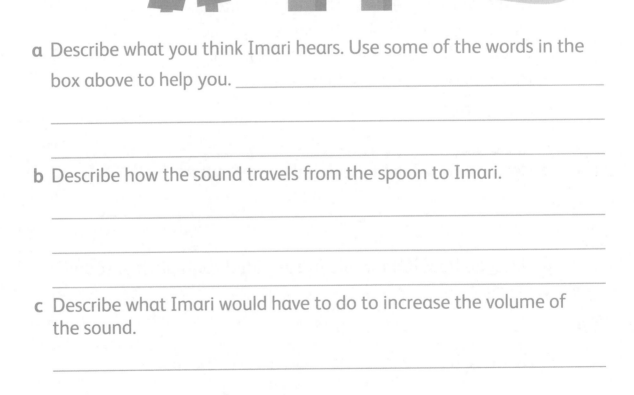

a Describe what you think Imari hears. Use some of the words in the box above to help you. _____

_____

b Describe how the sound travels from the spoon to Imari.

_____

_____

c Describe what Imari would have to do to increase the volume of the sound.

_____

_____

_____

Sounds in water

plastic tube   metal spoons   bowl and water

1 Sophia and Annie decided to find out if sound travels through water.

In the space below, draw a diagram to show Sophia and Annie what to do to with their equipment. Make sure that you add labels and notes to your diagram.

2 Describe what Sophia and Annie should do to find out if sound travels through water. Use some of these words to help you:

| water | hit | listen | hear | sound |
| travels | tap | spoons | tube | vibrations |

_____

_____

_____

_____

_____

_____

_____

_____

String telephones

1 In the space, design your own string telephone. Add labels and notes to your diagram.

2 Write a sentence to explain how you can hear someone talking, when using the string telephone.
Use some of these words:

| vibration | string | ear |
| passed | along | voice | cup |

_____

_____

3 Some learners tested different materials to find out which made the best string telephone. Here are their results.

a Write four sentences to explain how the learners carried out their test.

Material	How well they could hear
wool	*six words correctly*
string	*all words correctly*
rope	*three words correctly*
fabric	*no words correctly*

1 _____

_____

2 _____

_____

3 _____

_____

4 _____

_____

b Which material made the best string telephone? _____

c Why do you think that? _____

_____

Blocking sound

 Class 4 wanted to carry out a fair test to find out the answer to this question:

> Which is the best material for blocking sound?

They used these resources:

- ticking clock
- different fabrics (cotton wool, bubble wrap, paper)
- sound level meter.

Plan how Class 4 will carry out their fair test to find the best material for blocking sound.

a What could they do? _____

_____

b What could they change? _____

_____

c How could they make the test fair (what could they keep the same)?

_____

d What could they measure? _____

_____

e How could they record their results?

_____

f Now try the test yourself and record your results.

What are your conclusions? Which material is best for blocking sound?

_____

_____

_____

_____

Design a musical instrument

 In the space below, design your own musical instrument.
Make sure you annotate the diagram to show:

- the materials you have used
- how to change the volume
- how to change the pitch.

Sound challenge

 1 Here are lots of hexagons. Write words about sound in each hexagon. The challenge is to make sure that all the words in the hexagons link with the other hexagons around it. You could use some of the words in the box below. Some hexagons have been done for you.

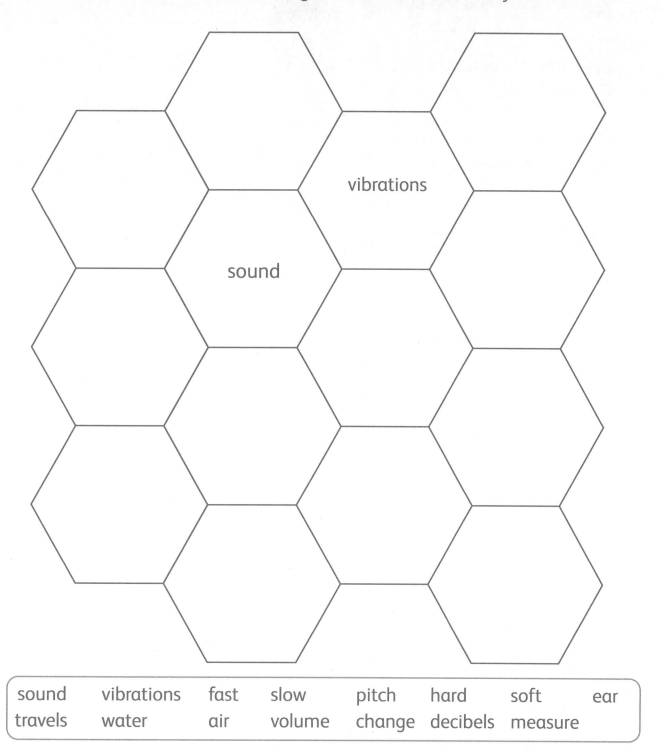

| sound | vibrations | fast | slow | pitch | hard | soft | ear |
| travels | water | air | volume | change | decibels | measure | |

Self-assessment

Unit 6 Sound

Learning objectives	😊	😐	☹️
I can use the word 'vibrate' to describe what happens when something makes a sound.			
I can describe how a sound travels to the ear.			
I know how to use a sound level meter.			
I can name the unit of measurement for sound.			
I can use the word 'vibration' to describe how to make a sound louder or softer.			
I know what 'pitch' means.			
I can explain how to change the pitch of a sound on a musical instrument.			
I can describe what happens when sound travels through different materials.			
I can name some materials that can block sound.			
I can name three ways that musical instruments can make a sound.			

😊 I understand this well.

😐 I understand this but need more practice.

☹️ I do not understand this yet.

I need more help with …

_____

_____

_____